✱ WHAT'S THEIR STORY? ✱

Galileo

Oxford University Press, 198 Madison Avenue, New York, New York 10016

Oxford New York
Athens Auckland Bangkok Bogotá Bombay
Buenos Aires Calcutta Cape Town Dar es Salaam
Delhi Florence Hong Kong Istanbul Karachi
Kuala Lumpur Madras Madrid Melbourne
Mexico City Nairobi Paris Singapore
Taipei Tokyo Toronto Warsaw

and associated companies in
Berlin Ibadan

Oxford is a trademark of Oxford University Press

Originally published by Oxford University Press UK in 1997.

ISBN 0-19-521405-6

1 3 5 7 9 10 8 6 4 2

Printed in Dubai by Oriental Press

✷ WHAT'S THEIR STORY? ✷

Galileo

SCIENTIST AND STARGAZER

JACQUELINE MITTON

Illustrated by Gerry Ball

OXFORD UNIVERSITY PRESS

Galileo Galilei grew up in Pisa, a town in what is now Italy. He was born in 1564, the same year that Shakespeare was born in England. His father, Vincenzo, was a musician, and Galileo had a talent for music too. He learned to play the lute and the organ.

Vincenzo always liked to try new things with his music. Galileo was like his father in many ways. He was a lively and energetic child, and he loved to explore and investigate everything around him. Vincenzo knew his son was clever, but he never dreamed he would be one of the world's greatest scientists.

Galileo had his first lessons from a tutor at home. He especially enjoyed drawing, poetry, and mathematics. When he was 10 years old, the family moved to the nearby city of Florence. There Galileo went to school at a monastery. But when Galileo showed interest in becoming a monk himself, his father was horrified and made him leave.

The Galilei family was not wealthy. Galileo would need to earn a living. Vincenzo decided his son should train to be a doctor at Pisa University. But Galileo hated it. He was sometimes mischievous. He kept arguing with his teachers and made himself very unpopular. His nickname was the Wrangler.

At Pisa, Galileo made an important scientific discovery when he was only 18 years old. It happened in the cathedral.

The priest was giving a very boring sermon. Instead of listening, Galileo watched a great chandelier swinging on a long chain. The swings were big at first, then got smaller. But all the swings—long and short—took exactly the same time. Galileo checked the time against the regular rhythm of his pulse. He was quick to realize the importance of his new discovery. Years later, he would use it to design a pendulum clock.

Galileo gave up medicine. He went back to Florence and studied mathematics. But he was still not qualified to do anything, and he could not get a job.

Galileo spent the next four years studying, writing, and experimenting at home. During this time he made friends with many influential people. He was a likable young man and had a lot of charm. Eventually one of Galileo's friends helped him to get a job teaching mathematics at Pisa University. But it went badly from the start. The other professors were very upset when Galileo mocked them. They were old-fashioned and teaching things that were wrong, Galileo said.

At that time, most people accepted ideas handed down from centuries before without question. But Galileo believed it was important to do experiments to prove that the ideas were true.

To make his point, Galileo did an experiment in public. In Pisa there was a leaning tower. It is still there today. From the top of the tower, Galileo dropped two different weights. The old belief was that the heavier one would land first. But Galileo thought that they would land together. He was right. The professors, students, and people of Pisa were amazed.

Galileo made many enemies at Pisa. They went to his lectures to hiss and boo. So after three years, Galileo moved away from his hometown to the University at Padua. Today Pisa and Padua are both in Italy. But when Galileo was alive, there was no such country as Italy. There were independent states, each with its own laws and ruler. Pisa was in Tuscany and Padua was in the Venetian Republic.

In Padua, people accepted new ideas much more easily than they did in Pisa. So Galileo did very well as a professor there. He settled down and had a family. He set up a workshop to make scientific instruments. The word soon spread about the fine things he made. Orders came in from all over Europe. Galileo spent 18 years in Padua. It was the happiest period of his life.

Galileo eventually grew restless, however. He wanted to live and work in a larger city and have more time for experiments. He hoped to go back to Florence, but he was not yet famous enough to get any job he chose.

At the university, Galileo taught astronomy as well as mathematics. He began questioning old beliefs about how the Sun and Earth move in space. Most people in Galileo's day believed Earth was the center of the universe. They thought that the Sun, the Moon, and the planets traveled around Earth in circles. But 20 years before Galileo was born, a man named Nicolas Copernicus had suggested a different theory. He said that Earth and the other planets travel around the Sun, and the Moon travels around Earth.

Galileo believed Copernicus was right. But not many people agreed. The leaders of the Catholic Church said Copernicus's theory went against the teaching of the Bible. In those days, the church authorities were very powerful. People who spoke against them were often tortured or killed. It was too risky for Galileo to say in public what he really thought.

When an exploding star blazed in the sky in 1604, Galileo became even more interested in astronomy.

In 1609, something happened that dramatically changed Galileo's life. He heard news of an amazing discovery. Dutch opticians had found out how to make a telescope. It was simply two lenses in a tube. But when you looked through it, distant objects seemed nearer!

Galileo set about making telescopes of his own. He soon made several, better than any of the Dutch ones.

Galileo arranged to show the governors of the city of Venice what they could see through a telescope. They went to the top of the bell tower in St. Mark's Square in Venice. The governors were very impressed by the wonderful new instrument. They could see a church in Padua 24 miles (38 kilometers) away and many other distant things. The telescope magnified everything nine times.

Galileo presented the telescope to the ruler of Venice, who was called the doge. In return he got a huge pay raise and was told he could keep his job in Padua for life. At last Galileo was really famous.

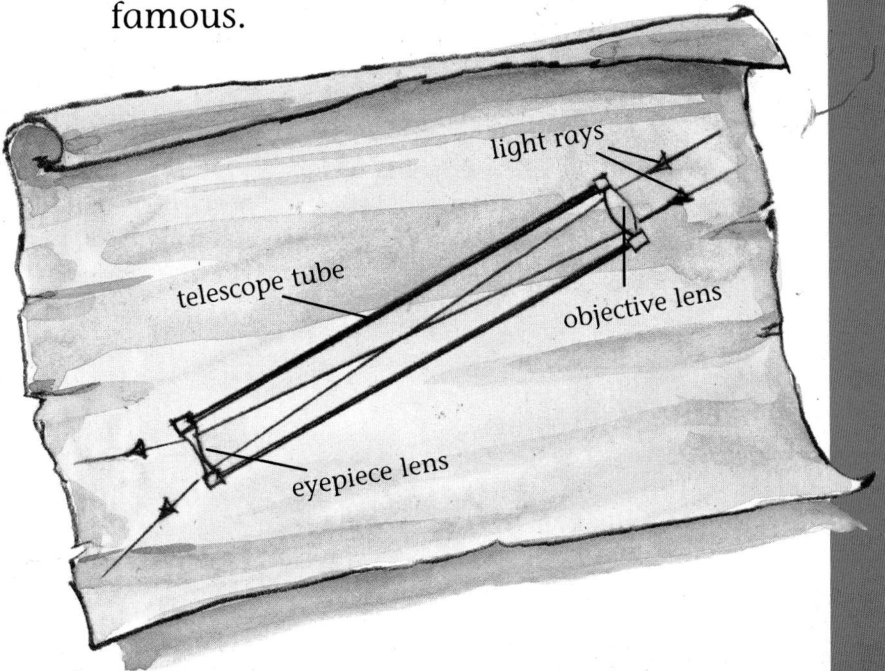

Galileo's telescopes were the very best anyone could buy. Orders poured in. Making accurate lenses was very difficult in those days, and Galileo set high standards. No telescope left his workshop unless Galileo thought it was good enough.

One clear night, he took his best telescope, one that could magnify 30 times, and turned it on the sky. What Galileo saw totally amazed him. Wherever he looked, there was something new—something you could not see by eye alone.

One of the things Galileo looked at was the Moon. The view thrilled him. People had believed for centuries that the Moon was perfectly smooth. Through his telescope, Galileo could clearly see huge mountains and craters. The Moon's surface was rough and uneven.

Galileo knew that these discoveries were dynamite! They would cause a scientific revolution. He could prove with his telescope that his own scientific ideas were right. How the old-fashioned professors would dislike him.

Everywhere in the sky, Galileo saw countless stars that were invisible without a telescope. Carefully, he noted his observations and made lots of sketches. It was all so exciting. What would he find next?

In January 1610 Galileo made one of his greatest discoveries. Observing the planet Jupiter, he noticed that it was in a line with three little stars. It might be a coincidence. But the next night the three stars were still there, although their pattern had changed.

For several nights Galileo watched Jupiter. On some nights he saw four stars instead of three. The pattern kept changing, and sometimes he could not see all of them, but there were definitely four little stars following Jupiter. There was only one explanation. Jupiter had moons going around it! When he could not see one of them, it was because it was behind Jupiter.

Galileo wanted to let everyone know about the wonders of the telescope. In just a few weeks, he wrote a book called *The Starry Messenger*. It created a sensation just as he had expected. But some people said he had made it all up and his enemies were jealous.

Galileo often upset people by the way he said things. He always said what he thought, and he had no time for anyone who disagreed with him. His friends warned him that if he went back to Tuscany, he would be in danger from his old enemies in the university and in the church. But Galileo was obstinate and ignored the warnings. He had a scheme for getting a job back home. He was really famous now, and he thought no one could touch him.

The family name of the grand duke of Tuscany was Medici, so Galileo called the moons of Jupiter he had found "the Medicean stars." (Today we call them the Galilean moons: Io, Europa, Ganymede, and Callisto.)

The plan worked. The grand duke was flattered and made Galileo his personal "Philosopher and Mathematician." So Galileo went to live in Florence.

But still many people did not believe Galileo's observations. So the next year, in 1611, he decided to go to Rome. There he would show the telescope to some of the most powerful and important scientists he knew.

The visit was a triumph for Galileo. With his outgoing personality and skill at lecturing, he captivated audiences with demonstrations of the incredible new instrument. Everywhere he went he was received as an honored guest. He had an audience with Pope Paul V, which was a great success. Galileo was overjoyed. At last he might be allowed to teach that Earth and the planets moved around the Sun, as Copernicus had said.

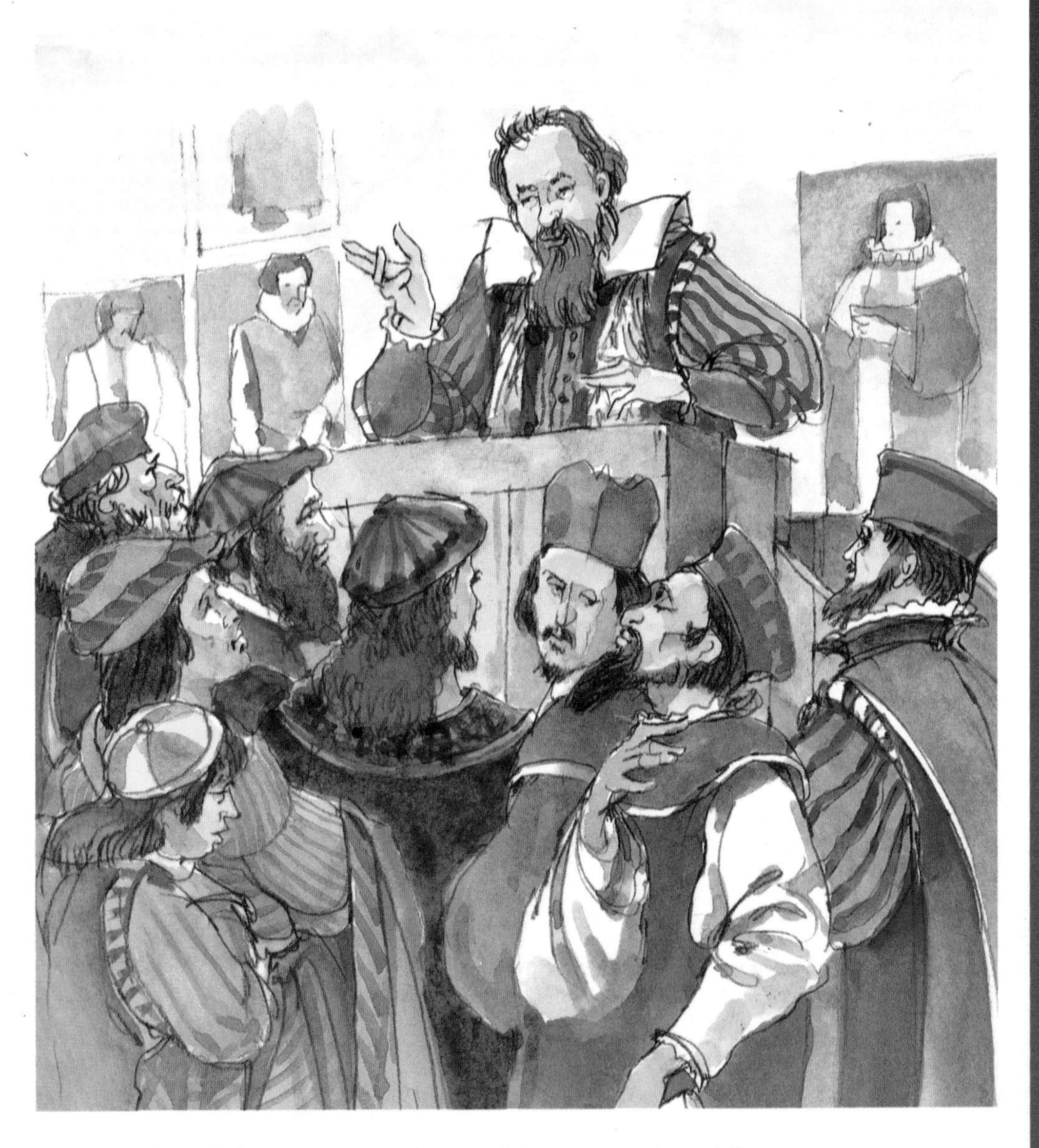

But he had forgotten about his enemies. They were very jealous because the grand duke paid him a huge salary and gave him special favors. In public, they criticized some of his scientific ideas. In private, they started to plot his downfall. Galileo found himself in more and more trouble as he tried to argue with his critics.

Things went from bad to worse for Galileo. In 1614, a priest gave a sermon condemning him and his ideas about Earth circling the Sun. Galileo's friends knew the situation was serious. They told him to keep quiet and lie low. Galileo was so worried that he became ill.

Even so, he would not accept his friends' advice. He went to Rome determined to argue his point of view. So the church authorities had to decide what to do about him. They declared that Earth must be the center of the universe. To believe Earth moved around the Sun was a serious crime against the church.

Galileo was summoned by a senior church official, Cardinal Bellarmine. He got a severe lecture and was told to go away and keep quiet or else. Disappointed, Galileo returned to Florence and worked on other things, at least for a while.

Eight years later Pope Paul V died. The new pope, Urban VIII, was an old friend of Galileo's. Galileo hoped the new pope would let him discuss Copernicus's theory again. He went to see the pope, and was delighted when he got his permission to write a book. But there was one condition. He was not to say that Earth really does go around the Sun. He must also say that the Sun could be going around Earth.

Galileo realized that he had no choice, so he agreed. He wrote a brilliant book. But he did not take the pope's warning seriously enough. His enemies saw their chance. They persuaded the pope that Galileo was making fun of him. The pope was furious and he ordered Galileo to Rome to be tried by a church court. It was a terrible shock. Galileo was 68 years old and unwell. He knew he could be tortured or put in prison if the trial went badly.

Galileo pleaded that he was too ill to go to Rome, but the pope was so angry he would not let him off. When Galileo felt better, the grand duke of Tuscany helped make his journey comfortable and arranged for him to be looked after in Rome.

The trial did not go well. Galileo still could not understand why so many people were against him. He was frightened of prison or an even worse punishment. But there was a way to avoid this.

Galileo agreed to read out the words that said he did not believe Earth goes around the Sun. In his heart he knew the words were untrue.

Later he went back to the court to hear his punishment. He had to wear a white robe to show he was sorry for offending the church. He knelt down and listened while he was sentenced to imprisonment.

In the end, it was not so bad for Galileo. He did not have to go to a real prison. He was sent back to his own house just outside Florence. But he was under strict orders to stay there quietly. His books were banned and visitors had to get special permission to see him.

Nothing could stop Galileo from thinking about science though! His imagination and curiosity in the world around him were as strong as ever. He was more than 70 years old, but he wrote a new and important book about how things move. A friend smuggled it out so it could be published in Holland. Galileo claimed he did not know how the book got out. Of course, that was not true! But he got away with it.

Sadly, Galileo went blind. Even so he kept thinking up new ideas. With the help of his son and friends, he worked to the very end. He died the month before his 78th birthday, leaving behind many important theories and discoveries that would help transform scientific thinking.

Important dates in Galileo's life

1564 Born in Pisa.
1574 Family moves to Florence.
1581 Begins studying medicine at Pisa University.
1582 Makes discovery about pendulums.
1582 Moves to Padua.
1585 Returns to Florence.
1589 Becomes lecturer at Pisa University.
1609 Learns of invention of telescope. Begins building telescopes of his own.
1610 Publishes *The Starry Messenger.*
1611 Shows telescope in Rome.
1616 Church condemns Copernicus's view of the universe.
1632 Publishes his book on astronomy. Pope orders him to Rome for trial.
1633 Sentenced to remain at home under arrest. Stays there for rest of his life
1642 Dies, aged 78.

Index

DATE			